Use Fractions to Multiply

John Carlin

Copyright © 2012 Author Name

All rights reserved.

ISBN:10:1501023179
ISBN-13:978-1501023170

CONTENTS

1	Introduction	1
2	A Basic Method	6
3	Starting Out	12
4	3x1 Digits	24
5	2x2 Digits	30
6	Synced Ratios	42
7	Cubing Fractions	51
8	Synced Ratios for 3 digits	55
9	2x3 Digits	58
10	2x4 Digits	66
11	3x3 digits	71

1. INTRODUCTION

I have spent a lot of time learning mental math shortcuts. As I studied, it became apparent that many of the books written were authored by people with savant like qualities. If you go to YouTube you can find videos of some of them doing mental math while hooked up to MRI machines. It turns out that most of them don't think about, visualize, or recall numbers like most people do. Once they get into the mental math mode, they see numbers in colors and shapes, and often they can't even tell you how they do their amazing computations. In their books, they detail methods that they don't use themselves because of their unique gifts in regards to calculating. They don't need or use these methods themselves.

There are few seminal works on mental math. The most helpful that I have seen are the Vedic Math books that have enjoyed a steady following. The father of the movement was Bharati Krsna Tirthaji Maharaja. He may have been one of the most brilliant minds to have lived in the last century. The story is that he studied ancient Vedic manuscripts, and rediscovered this incredible math system. This very learned man came to the United States, and lectured at some of the most prestigious universities about his methods. He authored a 16 volume work on the subject that was lost. By this time he was old, and in failing health. He was only able to replace his work with a single volume before he died. That single book became the source document, or Bible if you will, for Vedic Math.

This single book is insightful, and fascinating, but hardly replaces the 16 volumes that were lost. I have to believe that what was lost will forever remain submerged. Would it be as immense in size as what is under an iceberg's waterline? I tend to think so, judging from what was in the single surviving reconstructed volume.

Some people question this whole story. It reads like the plot of an old-time movie. The manuscript that everybody wants suddenly disappears. Did the Butler take it? Or was it the Gardner, or Chauffeur? The story would have made a great movie with perhaps a Peter Lorre type cast as the great Swami himself, and a Bogart type cast as the hardboiled detective. If you are a conspiracy theorist, the movie might resemble Oliver Stone's JFK classic. Would it turn out that the CIA, the FBI, or the Mafia were involved in not only stealing the manuscript, but possibly something more sinister?

I almost certainly dating myself with these obscure references to events and movies that happened, most likely before you were born. I may have dated references, but math is timeless. Remember that imagination in accounting or banking can cause an Enron debacle, or a financial meltdown. Infinite imagination in math however, equals infinite possibilities!

We have some math to learn, and I don't question that an Algebraic methodology applied to arithmetic would make all of us better at math. Others have come along too, and added to the body of work, most notably Kenneth R. Williams. If you are a savant, you don't need this book. This book will not turn you into one either. It will however, get you into the trenches with the numbers far deeper than the topical presentations that you normally see in books written by people telling you about methods they don't use themselves. It will make you better at fractions, math, and algebra. Mental math is largely about pattern recognition.

Everyone can get better at pattern recognition with practice. This book will point out patterns that you have never seen before. To a large extent, it is about changing your perceptions so that you can see the patterns that have been there all along. The traditional and very mechanical way that you learned to multiply using long form math doesn't do a thing for pattern recognition.

After a while, you will get very good at recognizing patterns and working with them. The patterns are draped over a easy-to-understand framework that mostly deals with related themes. The first theme is proportion, and that is the subject of this book. Math and proportion have a long history together. The slide rule uses proportion. Timekeeping devices like clocks use gears to help track and document proportion. Mechanical machine lathes use proportion to track and cut accurately. Before computers, the military used incredible mechanical devices to accurately compute the physics of gunnery and navigation.

Of course, it is difficult to learn mental math without touching upon some other things besides ratios and fractions. For one thing, some of the numbers have special properties, either by themselves or in conjunction with other numbers. The number one is an example. The identity property is used in mental math in ways that your grade school teachers never detailed. The commutative property of multiplication is another far underrated rule about multiplication. The numbers nine and eleven have some special properties as well, both by themselves and in conjunction with each other as we shall see, since 9 and 11 are one unit off from a base of ten. We then get into increments, and aliquot parts. All the while, there is Algebra, and yes, even Calculus lurking on the periphery. Don't ever let anyone convince you that learning anything beyond arithmetic these days is a waste of time. That is absolutely not true, and becoming more evident than ever in both our work and our everyday lives.

It used to be true that machinists used math more than anyone other than scientists and statisticians in their daily work. That is not the case anymore. Everyone has to up their game these days. Everything from professional baseball to marketing is subject to analysis that involves agility and comfort with numbers.

This book is about two digit and three digit multiplication. Anything that results in up to six digits as a product is included. So this book will also include methods for multiplying mixed digit numbers. The mixed digit multiplication that I detail is original material I have not seen in any other Vedic math books. The chapters on synchronized ratios are also original observations not pointed out in other books on the subject. One of the blessings of not being a savant is that I had to go into the trenches real deep with the numbers. I discovered things that savants would just have to repeat out of the existing literature. For example, you will learn some really cool things about threes digit by two digits multiplication; and you will learn they apply to four digits by two digits multiplication as well. You will see for yourself that, as the problems get bigger, the options for solving them expand exponentially. There is more detail here than a topical paragraph or two that just deals with the work lightly. Once you get into using fractions to multiply in detail, you will see that they are far more useful than you ever thought.

Three things are going to happen as you study. You are going to get better at fractions, and you are going to get more agile with numbers and multiplication, and you will be laying some groundwork to make Algebra your best subject. To repeat myself, everyone can get better at math by getting better at pattern recognition and by practicing. Hopefully, the practice will become play. Once that happens, you are on your way to becoming a math whiz. My best advice to you is to read this book in its entirety. If you get bogged down and it fails to hold your attention, determine to get through at least the two digits by two digits section. Yes! I am so into fractions that I am giving you the advice to at least read the first 1/3rd or ½ of the book. If you have taken, or are presently taking algebra, I urge you to read the whole book.

Use Fractions to Multiply

Practically everything in it can be related to algebra. There is no way that you could read this book and not become better at binomial algebra! Better yet, see my more advanced book.

https://www.amazon.com/dp/B009274SKK

2. A BASIC METHOD

If you wanted to multiply the numbers 39*21, the calculation could be done in three parts. The first part involves merely writing down the product of 3*2. The second part involves cross multiplying the 2 numbers. That calculation would be (3*1)+(9*2). The third component is the product of 9*1. The three components would be in serial order: 06/03+18/09, or 06/21/09. When you link these together you would get 819. With 3 digit multiplication, you merely split the 3 digits of each number into a 2 digit number and a one digit number, and then do the same 3 component calculation.

A selected example of three digit multiplication that involves splitting the three digits would be multiplying 309*291. In this case, I chose to split the 309 into 30 and 9. And I split the 291 into 29 and 1. The first component is calculated as 30*29. The second component is calculated as (29*9)+(30*1). The third component is merely 9*1. The three parts together are 870/291/09. The following visual will help you see how they are linked or added: 870/291/09=89,919

You link/add them together with an overlap, as though one component was ten times, or 100 times the next component, which in fact it is. When I put these numbers together, I get 89,919. That, in a nutshell, is the entire short form of this book. Put together two straight up calculations with the cross multiplication in the middle, and you have mentally multiplied any 2 or 3 digit number. The entire process is almost as mechanical and thoughtless as traditional long form multiplication.

I hope that you can see some of the advantages of using this method. For one thing, you can begin to immediately write the answer down from left to right, and link them one step at a time sequentially. Secondly, the concept of splitting the numbers into two parts makes them resemble fractions. Thinking of the numbers as fractions, and writing them a single line split like fractions, helps one method reinforce the other. This is the first step in changing your perception; it involves seeing the numbers somewhat differently. At this point, I don't advocate that you do any of these calculations in your head. Instead, use a notepad. Write the components down as you go from left to right so that you can see the progress of the calculation and the assembly of the components.

Write the answers down on a single line, as that is easier to link in your head than one over the other. For one thing, writing them down this way reinforces your agility with fractions. The links themselves resemble those fractions. Most of us are far more visual than anything else. Eventually, you will be able to switch over to the spelling bee approach that savants are famous for. Let us just set a modest goal for now of being able to do any one digit and two digit multiplication in our head by the time we are done with this book. To get practice linking the numbers, look at your digital watch occasionally through the day, and mentally slide the seconds one digit under the hours For example 19:59 would become the number 249.

These earlier examples were handpicked. Did you notice that the cross multiplications on these examples yielded a cross product that was exactly equal to the smaller number that was being multiplied? (Going forward, I will call cross multiplications by the term cross products, and the smaller number of the two being multiplied by the term the multiplier).

John Carlin

Look at the numbers before you start. You are looking for patterns. Once you spot the pattern, the math is greatly simplified. In the two previous examples, once the pattern was recognized, there was no cross product to calculate. I bet that you would have to look at these examples very closely to see what the pattern was that made these two calculations so simple. We will get to that later. The hardest part of mental multiplication would be getting the cross product. That is where the pattern recognition becomes so critical.

Surprisingly, the patterns are easy to learn and indeed follow a limited and consistent pattern themselves. The traditional grade school method of multiplication that you learned is all mechanical. In general, you just process the numbers one at a time. Grinding out the numbers without giving them much thought condemns you, in a sense, to reinventing the wheel every time you do a multiplication. Except for the occasional inadvertent discovery, there is no deliberate effort at a pattern recognition process. The three component process is not as likely to be as mechanical and thoughtless, because there are so many ways to simplify the process by looking at the numbers before you even start to do the problem.

If you take a moment and look at the numbers before you start processing, you will be much better served. Math power comes from developing a situational awareness of the numbers as they relate to each other. Initially, the time spent looking at the numbers will slow you down. Once you get that situational awareness, things change dramatically. All of the sudden you are processing numbers faster and enjoying the process far more than you ever imagined.

Resist the urge to just dive in and start crunching numbers! We both know that at some point you have to quit kicking the tires, and do the calculation, but you don't want to do it mindlessly either! With practice, you will easily recognize the patterns, and the math will be so simple and pleasant that you will be astounded. Usually with every problem there will be several ways to find a solution, and you will have the luxury of looking at

Use Fractions to Multiply

the problem and picking the way that you think is best to get an answer. This will be an enormous difference from the traditional long form multiplication that you learned in grade school. That method doesn't give you a lot of options.

We all know some math shortcuts. They are the result of our desire to cram and memorize by rote, as opposed to deeper learning. We all want the Cliffs Notes version. What usually happens with math shortcuts is that we get the little snippet but not the big picture. One often overlooked big picture item is that all the shortcuts have bi-directionality of a sort. That means that they work from left to right and from right to left. The results may look so very different that we learn to apply them in one direction but never in the other. What I'm saying is that the narrative may not be bidirectional, even though the problem is. Just as an example, consider for a moment the square of 13 and the square of 31. These are the same numbers just turned around. The problem and the narrative work in both directions.

One can easily see that bi-directionality is at work in the calculation. One answer is 169, while the other is 961. Not only are the numbers the same just turned around, but the answers are too. You could do the same with 12 squared and 21 squared and get 144 and 441 as answers. The numbers in the narrative about them falls nicely together. It would appear that you could make a rule out of this consistency. It would be real short and real simple. Unfortunately, the rule would be very short-lived.

Now consider 14 squared. Generally, we know that 14 squared is 196. What would 41 squared be? All of a sudden, 691 is not the correct answer! Let's look at the three components and see what happened. The three part answer for 14 squared is 01/08/16=196. The three part answer for 41 squared is 16/08/01=1681. The shortcut that you fixed in your mind as "just reverse the numbers in the answer," has to be modified to be "just reverse the order of the three components and put those together." The rules approach itself is mindless and can get you in trouble. Language allows you to change the narrative about the numbers, but it doesn't change the way that the numbers work. Part of the beauty of math is its immutability. Part of the beauty of language is its flexibility to describe immutability in different ways. Let each one rest on its own uniqueness!

If we square 15, we get components of 01/10/25=225. If we square 51, the reversed components are 25/10/01=2601. If you are using shortcuts to do mental math, you can automatically double the number of shortcuts available to you by realizing the fact of bi-directionality. Bi-directionality can be of two types. You can start on the left and move to the right, or work in the other direction. It also means that you can turn the number around, and still compute an answer just as easily. You could call bi-directionality an application of the commutative property of multiplication. Certainly when applied to two and three digit numbers, you can see that it is far more useful than the application to single digit multiplication that everyone learns in grade school and promptly dismisses with little thought other than to observe that it's lame. It's not something that you should learn and probably dismiss when it comes to mental math, because it gives you so much more problem-solving power.

Sometimes, it is easier to solve a specific multiplication by solving an entirely different problem. For example, solving 612 squared may be as easy or easier than solving 72 squared. In 12 squared, the last two components are each four. In 612 squared, the last two components are 144. No real cross product calculation is necessary if you recognize the proportion 6/12=1/2, (pattern recognition). The components of 612 squared would be 36/144/144. Put together, you would have 374,544 for an answer. If we put those same components together with the middle 144 slide all the way under the 36, since 7 is made up of 6+1 we get 5184, which is the square of 72 believe it or not. Don't focus on the math so much at this point, instead just absorb that it happens all the time in math and science that you sometimes solve problems easily by solving for something seemingly not related to the problem at hand.

Use Fractions to Multiply

Two physicists were riding in a hot air balloon, were blown off course sailing over a mountain trail, and were completely lost. They spotted a jogger running on the trail and they shouted, "Can you tell us where we are?" After a few minutes, the jogger yelled back "You're up in a balloon." One one the physicists said to the other, "Just our luck to run into a mathematician." How did they know the jogger was a mathematician?

He took a long time to answer; second, his answer was 100% correct and third, it was totally useless. :(

3. STARTING OUT

We are going to start out by looking at how you multiply a single digit number and a two digit number. We want to start out this way because you can't progress if you don't have this skill. It is the basic skill that you have to have to get on with learning more mental math. Let's say that you wanted to multiply 11*9. Eleven can be thought of as a fraction that equals 1. Think of it as 1/1*9. Now just multiply the leading 1 in 11 by 9, since we want to work from left to right whenever and wherever we can. You would think that the next step would be to multiply the second digit by 9 too. That is true; you can do that. Consider for a moment that since you know that 11 has a one to one relationship with everything you multiply it with, that means you know without doing the second multiplication that the second digit is going to be the same as the first digit. Thus 11*9=99, with just one multiplication. The one line link would look like 09/09 =99. (The zeros are not to be dismissed, they need to be there).

    The above example seems a little too easy, doesn't it? Okay, let's look at 84*9. Think of the 84 as 8/4. Now do that single multiplication on the left again. 8*9=72. What will the second number be without multiplying 4*9? It is going to be half of 72. Now we just have to link

Use Fractions to Multiply

72/36=756. This is a better approach, because you are linking off that first number anyway, and you are using the second number's relation to that first number. You will be better able to keep that first number in mind without the distraction of setting it aside, multiplying the second number, and then going back and linking it to the first number.

Using fractions to multiply single digit numbers is in fact as simple as it looks. Notice how 72/36 is the same fraction, or maintains the same ratio as 8/4. It is always this way; the ratio is preserved after you multiply. In the case of a 2/1 type of fraction, you can check your work with a simple glance at it. Observation is all that is needed

This will work pretty well for the easy fractions such as 1/2, 1/3, or 1/4. It won't work so good for some of the other fractions that we don't routinely work with. You always have the fall back position of being able to just do the multiplication. It doesn't appear that you gain that much here by using fractions. Remember, it is only single digit multiplication. You can't expect a miracle out of it. The big deal is that using fractions changes the emphasis so that you are looking at the links instead of the multiplication, and that opens up some other neat tricks. We will explore some of those later on.

I might also point out that 83*9 has an interesting property. Note that the digits in 83 add up to eleven. Anytime you multiply a number whose two digits add up to eleven, the component parts are reciprocals. The components in this case are 72 and 27, which is just 72 turned around. Look at 56*9, 74*9, and 65*9. The reciprocity holds right across the spectrum. You can have a lot of fun with this one.

Consider for a moment 803*9. In this case, the components are still 72 and 27, only now they are put together as 7227 instead of 747. 902*9 would be 8118, while 92*9 would be 828. There is an aliquot angle to this also. Since 4.5 is half of 9, it should follow that 83*45 would equal 72/27 divided in half. That would result in 36/135 being linked. That answer would be 3735. In the 84*9 example, we started out with the fact that you could solve for 83*9, which is an easy 747. Now just add nine to that to get 756. What would you do if the problem were to find 82*9? If you said 747 minus nine is 738, you are right on the mark!

There is a lot of leverage that you can get when eleven and nine are involved. Consider 838x9. In this case, if you split the number into 83/8 or 8/38, the digit sum for each part is eleven. The vertical calculations begin and end with the same number, which in this case is 72. Now the middle part is that same number turned around. The expansion becomes 72/27/72= 7542.

Forget all the math, just note that the digits add up to eleven no matter how you split them. Your multiplier is nine, so the components are 8x9=72 for the beginning and end. Stick 27 in the middle since you know your stuff. and the answer is reduced to a mere assembly of the three modules. Try 656x9. We do this with one calculation once you see the digit sum eleven. Six times nine equals 54. Now we have everything that we need to do the math. The components are 54/45/54. The answer is: 5904.

If we were multiplying a two digit number with a digit sum of eleven by 99, what would happen? Let's try 83x99. The middle component will always be 99. Here is how it would work. You still have the two reciprocal numbers as your vertical calculations. The cross product is the sum of 8+3 multiplied by 9. Your cross product is always going to be 99, since the digit sum is always 11. The components to put together are 72/99/27=8217. Try it with 38x99, 47x99, and 56x99. The methodology is always the same.

If you needed to multiply 838x99, this same leverage would come into play. Now you would have 72 and 72 as your vertically derived components. The cross product becomes a cousin to 99, it becomes 1089 derived by linking 99/99 to get 1089. Try this with 474x99, 656x99, 383x99, or 292x99. One calculation --the rest is just putting the pieces together. The logical question I hope that you ask yourself would be what if the digits came up to ten, and not eleven? What if they added up to twelve instead of eleven? Do some of the

Use Fractions to Multiply

problems as though they added up to a digit sum of eleven, then see for yourself what the correction would be if you were over or under by a single unit.

Also, remember that many of the single digit multiplications you would never do as fractions. For example, if you are multiplying a two digit number by a multiplier of 1, or 2, perhaps 3, and 4, or 5, you wouldn't be bothered with splitting it into a fraction. Because of the identity property, multiplying by one just gives you back the number itself. Multiplying by two just means that you double the number you have. Multiplying by three, you just triple it. With four, you could double twice. I'm sure that at this point, you know you can multiply any number by five by tacking a zero onto the end of multiplicand, and dividing the net result by one-half.

If the multiplier is the same as one of the two digits, the multiplication becomes a matter of squaring the digit that appears twice and linking it to a number that is a multiple of the multiplier. For example, with 82*8 you would look at the 8 and think 64. Now it is just a matter of linking the 64 to some number that is a multiple of eight, in this case 1/4th of 64. In this case, it would be 64/16=656. The math is especially easy if one of the two digits is one, and the other one is the same as the multiplier. For example, 81*8 would result in eight squared and eight as the digits to link up. The fraction would look like 64/08=648. It is merely eight squared with eight tacked onto the end of it.

If the ratio of the numbers being multiplied is close to being a whole number, that is not so difficult to deal with as you might at first think. For example, 67*8 can be thought of as 6/7ths. You are 1/7th away from a whole number. You could multiply the 6*8 to get 48 and link it with 48, since that would result in a 1:1 ratio. 48/48= 528. This is not the right answer it needs to be corrected by adding 8 to 528 to get 536. The one increment you needed was just to add the multiplier back into the whole number you linked so easily. If the problem had been 76*8, you would have linked 56/56=616 and subtracted 8 to get 608, since you were stepping down by an increment of one. Of course if you were stepping up or down by two, that being the difference between the two numbers, you would add or subtract by two times the multiplier. The method is quite simple and effective. It is also a classic case of solving a problem by solving a different problem and incrementally adjusting that answer.

As you might guess, you can use this same incremental approach in different scenarios. For example, if you are going to multiply (38*6) you could think of 3/8 as being one unit away from 4/8. 4/8*6 would be a real easy 1:2 ratio to work with. If we were to link 4/8*6, we would have to link 24/48=288. The part of this that we have to adjust downward is the first part, since we are solving for one unit less than the four in 4/8. We are also working in the tens part of the link rather than the units part of the link. That means that we have to subtract 60 in this case from 288 to get our final answer of 228. This approach combines aliquot parts with the use of increments. The same methods can be applied to other mental math problem-solving approaches. Get used to seeing these two methods used together. Almost instantly, you zero in on a number very close to the right answer. Then you just clean it up a bit with a simple adjustment.

If you think about it for a moment, what you're really doing in these cases is looking at the difference between the two numbers. There is another narrative that we can put with this method. It doesn't change the numbers or the way that they work together, but it may change how you think about this problem. Have you ever heard of checking your math work by casting out nines?

What you do is add the digits in the two numbers you are multiplying. Multiply those two digit sums together, and compared them to the digit sum of your answer. In the case of 67*8, notice how this works. The two numbers added together, six and seven total to 13. You can subtract nine from 13 to get four, or you can just add the 13 together one more time to get four. Now we take the four and multiply it by eight to get 32. Our answer was 48/56=536. Again, we can add these digits to

Use Fractions to Multiply

get the single digit of 5, or just throw away the 3 and 6 since they add up to nine. Either way, we get five again, since 3+2=5. We say casting out nines is a good way to check your work. Nine in this case is like zero. It doesn't affect your answer.

When we're solving for 77 times eight, we're solving for a digit sum of 14 instead of 13. That change in one unit in the digit sum translates to a change of eight in the answer. That change of one unit also happens to be the difference between six and seven multiplied by your multiplier. So stepping down or up by one unit simply means stepping down or up by the amount of your multiplier in the answer. In the case of 67*8, we can take the number 48 and just step up or down eight units to get the second linking component. Nothing could be easier if you are stepping down from 48. The next component is 40. If you are stepping up, the next component would be 56.

The same principle of stepping down, or up can be applied when you are looking at any number that you want to adjust by a single or double amount. In the 3/8*6 example, we noted that we were 1/8 away from 4/8. The difference between a three and four is one for 4/8*6= 24/48. Now we are adjusting the 24 downwards by six. Therefore, we have to link 18/48= 228. In general, it is easier to adjust the link than to adjust the total. Adjusting the digit sum by one or two units means adjusting one of the two links by one or two times your multiplier. The direction of the adjustment is additive or subtractive depending on whether you're adjusting upwards or downwards.

There is another approach that you can use that works off the digit sum and is especially useful when the spread between the digits is rather wide. Let's look at (92)*8. If you add the two digits 9 and 2, you get 11. Now split 11 into a fraction. The fraction made out of the digit sum can now be multiplied by the multiplier, and the links joined as in normal single line multiplication. Now you would be linking together 08/08=88. Go back at this point and multiply the smaller number in the original problem by the multiplier. So (2*8)=16. That is half the answer. The other half is 88-16=72. Now just link them together: 72/16=736. This is not the most effective way to multiply, but it demonstrates how digit sums can be used to not only check your work, but actually be used to multiply.

The above method gets you outside the box, and may have some problem solving usefulness when you get into algebra. If you are already taking the subject, contemplate for a moment how digit sums may help you factor an equation, or at least check your work. It is interesting how loading that digit sum onto the front of the link gave you 88 to work with, so you now had a number that was usable to solve the problem. If you added it the traditional way, you would have had 16 to work with and that would not have been of much value for solving the problem. The digit sum of 88 adds to 16 though, so once again we see how the form shapes what we see or don't see.

Nine as a multiplier is particularly interesting. To me, it is the verb of numbers. There is always a resultant action from nine, and it modifies every number it is associated with. Every multiple of nine has a digit sum of nine. Add the digits in 81, 72, 63, and 54. They all add to nine don't they? You know that you can check the accuracy of your multiplications by casting out nines. Every whole number or decimal, no matter how long it is, can be condensed down into a single digit. The single digit is called the number's digit sum. To get it, you simply add up all of the digits in the given number. If this sum is larger than one digit, you add its digits together too. You keep doing this until the result is a single digit.

In the case of multiplication, you know that you can use this fact to check your work. For example, 25*25=625. The digits in 25 add up to 7. 7*7=49 that digit sum is 4 because the number 9 in this case is like a zero. It has no effect on the answer. Now let's add up 6+2+5=8+5=13=4. Since this result is the same as what we got when we multiplied 7x7 and added the digits, we probably have the right answer for the multiplication. I say probably because some other combination of inadvertent numbers that also added up to four could

Use Fractions to Multiply

have given us a false/positive test result. The odds of this are unlikely, but could happen.

Earlier, we went through scenarios where we adjusted a fraction that was close to being a whole number by computing the ratio for a whole number and adjusting it by the multiplier. Just as we took a fraction like 7/2 and turned it into 6/12, we could just as easily have turned into 8/1. You can't get a simpler fraction to work with than one that is already a whole number. If you wanted to multiply 72*6, you could, besides turning it into (6/12)*6, turn it into (8/1)*6. In this case, you have to link 48/06=486. Your adjustment in this case would be subtractive, and it is always going to be a multiple of nine times your multiplier. In short, you would subtract (6*9)=54 from 486 to get 432.

As a practical matter, it's not likely that you would convert something with a numerator or denominator into a whole number or it's reciprocal to solve a single digit multiplication. Generally, these problems are so easy that there is no need to get especially creative to do these multiplications. I point this one out to make the point that adding ahead or borrowing back does not change the digit sum; and since nine is outcome neutral in the digit sum world, it follows that your adjustment would be a multiple of nine. Just for practice, convert 2/8 into 1/9 and multiply it by 6, 7, and 8. Compute the adjustment, and compare your answer to one you got by another means. Did you also notice that the digit sums for 114, 133, and 152 all add up to 6, 7, and 8 respectively? Everything falls into a pattern that is easy to discern.

Anything that you multiply by nine will have a digit sum of nine. Another characteristic of nine is that it's closely related to 11. For example, anytime you are multiplying a two digit number that adds to 11 by 9, the link will be the reciprocal of one of the multiplications. For example, 74*9 will have links that are reciprocals of each other. 9*7=63. The other link is 36. In single-line notation, this would look like 63/36=666. Try multiplying 92*9, 83*9, 74*9, 65*9, 56*9, 47*9, 38*9, and 29*9. Also, if you think about it, a three digit number that has a digit sum of 11 multiplied by nine will have the reciprocal portion spread out in an exact ratio of the numbers to themselves. For example, (812)*9 you could do with one multiplication of 9*8=72. The reciprocal is 27, and it is split into two parts, one of which is twice the other part. So that would give you links of 72/09/18=7308. What would happen if the digit sum was 22 and your multiplier was 9? Play with it a little bit, and see for yourself. It is these little nuances that separate the craftsman from the hack.

Just visualize the left most multiplication and link it to its compliment or reciprocal so to speak. If you think about it, linking the reciprocals by adding them together is not even necessary, although it is a good way to remember the rule. The odd coefficients, (the 1st, and last digits), are always going to be equal, so if you double them and check the digit sum they present you can know the middle or even coefficient without much effort, since you know the digits have to add up to nine.

Any two digit number that is a multiple of 11 when multiplied by 9 will have nine as the middle digit. The first digit and the last digit will be the product of the multiplication with nine stuck in the middle. For example, 88x9 would result in an answer of 72/72=792. 66x9 would result in 54/54= 594. The nine stuck in the middle is always a placeholder in the first and last digits add to nine.

Any two digit number that ascends by one when multiplied by nine will have zero as a middle digit. For example, 89*9= 72/81=801. Multiply 12*9, 23*9, 34*9, 45*9, 56*9, 67*9, or 78*9 and review the answers. Notice that in this case your answer is always the product of the last digit (the larger digit) split apart with a zero inserted in the middle as a placeholder of sorts, in the course those two digits add up to nine.

What happens when the digits descend in steps of one and your multiplier is nine? Good question! Glad you asked! The middle digit will then always be eight, and the first and last digits will be complements of 10. It will work out that the first and last digit will be the product of the first digit multiplied by 9 with 1 added to the last

Use Fractions to Multiply

digit. Try multiplying 21*9, 32*9, 43*9, 54*9, 65*9, 76*9, 87*9, and 98*9. The multiples go up by 11, the middle digit is 9-1=8, and the first and last digits sum up to 9+1=10. It all follows a pattern. Thus, 98*9= 81/72=882. Eight is in the middle between 9*9=81+1=82.

What happens when the steps ascend or descend two at a time and your multiplier is nine? Once you get one part of the puzzle, the rest becomes quite easy. If steps were ascending, the middle digit would always be 9+2=11. If you were descending, the middle digit would always be 9-2=7. Let's look at 13*9, 24*9, 35*9, 46*9, 57*9, 68*9, and 79*9. In every case, the middle digit is always 1. The first and last digits will always add to eight. That way, we preserve the digit sum of nine as 1+8=9. The first digit will always be the first digit of the number that you are multiplying. The last digit will be 8 minus that number.

In the case of descending order in steps of two, the middle digit will always be 9-2 = 7. That means, of course, that the first and last digits have to add up to 11. The last digit in this case will be the tens complement of the last digit that you are multiplying. The first digit will be 11 minus that number. Try these out: 31*9, 42*9, 53*9, 64*9, 75*9, 86*9, and 97*9. We can go on from here just as we could have gone on in the case of multiplying two digits that are almost a fraction of 1. In the case of multiplying by 9 or 11, these adjustments sort of degenerate into some rules that you may find easy to use. Personally, I am not a fan of these because they skip over the reasoning that goes into them. That, of course, is an injustice to the incipient math learner. I will present them here and leave it to you to use them or not.

1. To multiply any two digit number by nine, start on the left and subtract that first digit from the second digit. Then subtract the second digit from that first number with a zero tacked onto it. For example, 93*9 would be 840-3-837. In essence, the last digit is the tens' compliment of the last digit of the multiplicand. The middle digit 84 is always reduced by one unit since you are subtracting.

2.To multiply any two digit number by 11, the process is reversed. You just add the first digit to the second digit, and tack the last digit back onto your answer. This is exactly what you are doing if you think about it when you link two numbers being multiplied by a 1:1 ratio. 93*11 would be 93/93=1023.

3.The equivalent in Algebra to nine is the expression (x-1), where x=10. Anytime you multiply any algebraic expression by x-1, the three coefficients sum up to 0. If they do, then x-1 is a factor of the equation. With the number nine, it is not zero that the coefficients sum to, it is nine itself. The algebraic expression (x+1), where x=10 is equivalent to the number 11. You can always tell if x+1 is a factor in an equation, because the middle or even coefficient equals the sum of the odd coefficients.

4.You can solve any two digit by one digit multiplication by adding the two digits together and multiplying each digit by your single digit multiplier. Link those two numbers together, just as you would one line link any multiplication. Now subtract the result of one of the two multiplications in the original problem to get the second number that you need to link to get an answer. For example, in the problem 67*8 6+7 equals 13. 1/3*8=8/24=104. Subtract 6*8=48 from 104 to get 56. 48/56 are the two numbers you have to link to get your final answer of 536. This works especially well if the numbers are far apart. If the numbers are close together, you want to take the opposite tack and look at the difference between the two digits and adjust by that.

We have covered the basics of single digit multiplication fairly extensively. We also had some practice with linking numbers in a single line. We have seen how fractions tie into links and multiplication itself. We have started down the road to two digit multiplication by looking at how to multiply by 11. We have also investigated the number nine and its unique properties, including digit sums. From here, we probably want to look at three digit by one digit multiplications.

What is the difference between philosophy and math?

Use Fractions to Multiply

**Philosophy is about objectives with no
rules, math is about rules with no objectives :)**

4. 3X1 DIGITS

Essentially, 3 x 1 multiplication is the same as 2 x 1 multiplication with one more iteration, or step in the process. If the numbers being multiplied were all multiples of 11, they all would have a one-to-one relationship with each other. (777)*6 would involve linking 42/42/42. 4,662 would be your answer. One multiplication two links would be all that was involved. The math would be real simple also if the fractions fell right in ratio. For example, (842)*7 would result in links of 56/28/14= 5,894. Notice how the links started out with 56 and halved with each multiplication. Not all the numbers will fall in line like this, but the addition of one more digit in your multiplicand exponentially increases the number of options that may present themselves in doing the multiplication.

In short, the multiplication isn't complicated by the extra digit, and the opportunities for finding solutions are multiplied. Now we can bring in the usual suspects and take a look at them. The literal numerical line up. You will easily be able to identify the characters involved. If the numbers to be multiplied begin and end with the same digit, you know that you're just going to be replicating one of the multiplications. Even better if the

Use Fractions to Multiply

middle digit is fractionally related to the first or last digit. For example, 848*9 can be broken down on sight to 72/36/72=7632. If you have a one anywhere in the multiplicand the problem is simplified.

You can name the relationships if you want with names such as twins, little brother, big sister, or half sister. There could even be cousins, which would be multiples of another number, second cousins would be two multiples removed. To me, nine and one are cousins, each once removed from ten. Borrowing back, for example turns 72 into 6/12, and is creating a step sister or brother. I am sure that you get the picture at this point. In the genetic modification world, scientists are doing the same thing with gene sequences. They are studying patterns and relationships,the starting point for which may as well be mental math, fractions, and ratios.

If you don't like the fractional relationship of one of the digits to the other, you now have the option of relating it to a different fraction. In the problem (943)*6, the last digit matches easily with the first digit. On this one, I started on the right and figured 18, added six more since the difference between the numbers was one unit, then I tripled the 18 to get 54. The components were then 54/24/18=5,658. The last digit was the one that related to everything, so I started there and went with it. I started at the branch end of the family tree so to speak because it was easier.

It also can happen that the three digits can be split into a single fraction, and you proceed from there. For example, 824 could be split into 8/24. If you were multiplying that by 6, for example, you could just do this calculation all in two gulps, as 48/144= 4944.

This would be the point where we do some exercises as opposed to experiments. With these, you are getting practice that will make single/double digit multiplication seem very easy, and you can get a feel for how your options are multiplied by the increased digits. The chances to be a creative problem solver are greatly increased. Remember to walk around the problem before you dive into finding a solution, try to start on the left first, and think fractions if possible. The answers will be in link form on the following page.

Use Fractions to Multiply

Exercises

1. 789*6
2. 456*7
3. 123*8
4. 741*9
5. 852*6
6. 963*7
7. 147*9
8. 358*8
9. 369*7
10. 159*6
11. 627*3
12. 843*7
13. 623*8
14. 753*9
15. 829*7
16. 643*3
17. 792*7
18. 849*4
19. 463*7
20. 925*8
21. 357*4
22. 989*6
23. 741*6
24. 157*9
25. 931*6
26. 583*4
27. 367*8

28. One hundred percent markups on products are not that uncommon on certain products. The extended fraction 01/02/04 could be thought of as a representation of that markup scenario. The middle fraction would be the wholesale price, the first fraction would be manufactured price, the final number would be the retail price. If such a markup system were placed upon IV drip bags used by hospitals, what would be the difference between the manufactured price and the final price charged to the insurance company on a bag that cost 9 dollars from the manufacturer?

Answers

1. 42/48/54
2. 28/35/42
3. 08/16/24
4. 63/36/09
5. 48/30/12
6. 63/42/21
7. 09/36/63
8. 24/40/64
9. 21/42/63
10. 06/30/54
11. 18/06/21
12. 56/28/21
13. 48/16/24
14. 63/45/27
15. 56/14/63
16. 18/12/09
17. 49/63/14
18. 32/16/36
19. 28/42/21
20. 72/16/40
21. 12/20/28
22. 54/48/54
23. 42/24/06

Use Fractions to Multiply

24.   09/45/63

25.   54/18/06

26.   20/32/12

27.   24/48/56

28.   The markup chain, not to be confused with a Markov Chain, would take the form of 09/18/36. The big point here is that these links can be used to easily explain many things. In this case, they may explain why cutting out the middleman can reduce costs, increase profits, or do a little of both. Alternatively, they may explain why things sometimes have outrageous markups on them. This is particularly so if the item is a "got to have item" that is not easily done without, as in health care items. See also elasticity of demand as a term used to describe just how "got to have" an item may be.

A mathematician and an engineer are on desert island. They find two palm trees with one coconut each. The engineer climbs up one tree, gets the coconut, eats. The mathematician climbs up the other tree, gets the coconut, climbs the other tree, and puts it there saying: "Now we've reduced it to a problem we know how to solve." :(

.

5. 2X2 DIGITS

One of the most commonly known shortcuts is about how to square any two digit number that ends in 5. For example, if you wanted to square 15, or 25, or 35, just multiply the first digit by one more than its value, and tack on 25 to finish the calculation. The problem 25 squared could be calculated as 3*2, with 25 slapped onto the end of that calculation. That would give you 625 for an answer. 35 squared would be 1,225. 45 squared would be 2,025. It's about as simple as can be.

Now let's look a little closer at the components to this calculation, and see if we can expand the rule and determine if bi-directionality is involved. 35*35 would have the components of 9/30/25. 9/30 becomes 12, which is really three times four. It is easy to see how the rule works in this manner. Notice that the last two digits of this calculation sum to 10. Or to put it another way, notice that the tens'

Use Fractions to Multiply

complement of five is five. Is that significant? You could say that these two numbers have the same tens' base and the units' digits add to 10 or are tens' complements of each other. Let's look at 31*39. Notice that the tens' base is the same and the units still consist of numbers that add to 10. What would our cross product or even coefficient be in this case?

It looks like this would break into 9/30/9. That means that our answer would be 1,209. The 9/30 again becomes 12, or three times four. It appears that the rule can be expanded to include all numbers that have a similar base and units' digits that consist of a number and its tens' complement. Just out of curiosity. can we apply this rule bi-directionally? What does 52*52 equal? The components would be 25/20/4 equals 2,704.   The middle coefficient in this calculation is also 20. It makes sense that the cross product would stay the same, but the narrative has to change somewhat, since 6*5 would be 30 and our middle coefficient is 20. The fact that the narrative is not valid does not mean that the calculation is not bidirectional. Now the cross product is equal to 10 times one of the second digits. 52 squared has a cross product of 20. 53 squared has a cross product of 30. If we took 31*39 and tested for bi-directionality, we would be looking at 93*13. The cross product is still going to be 30.

You can multiply two numbers together by taking the average, and squaring that number. then subtract the square of the difference between the two numbers. This is one of the most basic symmetry examples that you can come up with because by definition the difference between the two numbers is equal. Average in this case means equidistant from each other. For example, 9*1 could be computed by noting that they add to 10 and the average of the two numbers is five (each number being four units from five). Square the five to get 25, and subtract the square of four (equals 16) to get nine. It's a left-handed way to get 9*1, but it works! You can apply this to other numbers as well though. 37*43 would have an average of 40. 40 squared equals 1,600 and three squared equals nine. So subtract nine from 1,600 to get 1,591 as your answer. This method is especially useful when the average of the two numbers is a number that ends in zero or double zero. Think of 49*51 as 2,500-1=2,499.

I would point out that your mental math calculations will be greatly sped up by knowing all the squares of the two digit numbers 11 through 99. One way to square those numbers in your head is to multiply the first digit by the second digit and double it to get your cross product. Tack that cross product onto the square of the first digit, and tack the square of the last digit onto the last digit of the cross product. For example 67 squared would be 36/84/49=4,489, note: 84=6*7*2.

There is a variation on this that may be useful at times. To get the cross product of a square, you can do your first vertical calculation, and get that first link. Now multiply that lead digit again by whatever the difference is between your digits added or subtracted from the second digit. In the previous example, we squared 67. That first vertical calculation would result in the number 36. Now take this six again and multiply. This time by eight, as that is the difference between six and seven added to the second digit. 6*8 is 48. If you add 36 and 48, you get 84, which is your cross product.

If the number were descending by one unit, then the problem would be to find the product of 65 squared. In this case, you would multiply six times four to get 24. Now add that to 36 to get 60. 60 is in fact the cross product. If the difference were two units between the digits, then add two to your second digit, or subtract two if you were descending. Then do the multiplication and add that result to that first multiplication to get the cross product.

Granted, this is not a practical way to do the math. It is a reasonable way to check your work, however. The main reason that I point this observation out is that it resembles another method for mental math that is quite effective and useful. Just as you can adjust a two digit number that ascends or descends when you are multiplying by a single digit, you can adjust a square in a similar fashion.

To multiply any two digit number that has the same base and ends in 5, just multiply that base by one unit more than its value and tack on 25 to that product. 75 squared, for example, would be 8*7 equals 56, with 25 tacked onto it to give you 5,625. 61*69 would be 7*6=42+tack on 9*1=09 to get 4,209.

There is a whole class of problems that have the same base. Every time you square a number, you're working with numbers that have the same base! There is a fractional approach that we can use on these problems too. You could think of a number like 14 squared as a fraction of one fourth or the whole number four. You can get the cross product by just adding the fraction, and multiplying by that times the product of the numerators so to speak. The fractions 1/4+1/4 equals 2/4 or one half. Now you can use the square of either the numerator or the denominator of your fraction to get the cross product of 8. The whole number four added to the number four equals eight. Now multiply that by the square of the lead digits. 1*1*8=8.

In another example, 84*84 could be split into the fraction 8/4=2. Now just add the fractions together to get 16/4=4, which is the multiplier for your cross product. This one works out very sweetly. The link starts out with 64, the next one is 64, and the final link is 16. 64/64/16=7,056. The method is especially easy when the fractions add up to the whole number one. 24 squared equals 2/4+2/4=1. 4*4=16, and so does the cross product, you don't even have to think about it. So with one calculation and knowing the relationship is 1:1, you have 2/3rds of the problem solved. Note how you have an option when making your fractions. In the case of 8/4, you can read it as 16/4=4, or 88/=1. In one case, you play off the first vertical multiplication, while in the other case you are playing off the second vertical multiplication. The cross product becomes 4*16, or 64*1. Either way, the cross product is 64.

To take this a step further, the logic still works if the fraction adds to one and the base is equal but the numerators are not the same (you are not squaring the number). You still have a cross product that is equal to one times the lead digit(s) or the following digit(s). 13x23, for

Use Fractions to Multiply

example, has a common denominator of three so to speak, so if you add the numerators in this case one and two you get 3/3=1. In this case too, the cross product is equal to the square of the common denominator. 3*3=9, so your components on this are 2/9/9=299. 78*18= 7/64/64=1,404, since the fractions 7/8+1/8=8/8 or 1.

The example of 13*23 is almost a reciprocal. Now may be a good time to mention that reciprocals have unique properties and are very easy to work with too. The pure form reciprocal problem could be something like 27*72. Any time that you multiply a number by its reciprocal, you will get an expansion where the first and last digit are equal and the even coefficient is equal to the sum of the squares of the two digits. A big mouthful that simply means, for example, that 27*72=(2*7)/(2*2)+(7*7)/(2*7). That would be 14/53/14= 1,944. The whole calculation could be done using either number 27 or 72, and then start multiplying 2*7=14. You know the last component is 14 also, and the middle component simply 2 squared +7 squared. In algebra, if you saw the coefficients 14/53/14, you might at first think that since the first and last components were equal maybe (x+1) might be one of the factors. Any number that is 11 or a multiple of 11 is a reciprocal where the difference between the digits is 0.

The multiplication of 77*77, for example, would have the components 49/98/49, and the second term, besides being twice the first or last term, is equal to the squares of the digits added together. In the problem of 27*72, since 14+14 equals 28 and the second coefficient is not 28, we know (x+1) is not a factor. So what would be another possibility? The middle coefficient is equal to the sum of the squares of the digits. We can quickly find the factors by finding the difference between 28 and 53 and taking its square root. In this case, the difference is 25 and square root of that is five. Now we know that we are looking for two numbers that multiplied together equal 14 and are five units apart. The puzzle is basically solved! The only possibilities are 2 and 7. So (2x+7)*(7x+2) are the factors of this expression.

Once you see that the two fractions add to one, you have instant knowledge of what the cross product is and combined components are. For example, with 14*34, you would instantly recognize the last 2 components as 16/16. The combination is as simple as the sum of the digits of 16, which is 7. So now you are mostly done: 3/176=476. This is an easy averaging problem also, as 24 squared equals 576. Now just take 100 away from the answer.

Guess what? This method works even when the denominators, so to speak, are not the same. A two digit version of this could be finding the product of 39*46. 3/9 is 1/3rd. 4/6 is 2/3rds. The sum of the two fractions is one. The components are 12/54/54. The link of the last two numbers is equal to the sum of the digits in the last component or the cross product. 12/594=1,794. Notice how that 594 component contains the sum of 5&4? What could be easier? This is a very easy calculation. The pattern and link are quite easy to learn and work with.

Use Fractions to Multiply

You can have ratios greater than one. All that means is that the cross product will be greater than the denominator. For example, 81*92 would have ratios of 16:2, and 9:2 added together you have 25/2=12.5. The cross product is therefore 1*2*12.5=25. For example, in 91*82, the respective fractions are the whole number nine and 8/2=4. Together they add up to 13. The denominator is 2*1=2, and the cross product is 13*2=26. The components then are 72/26/2=7,462.

For another example, look at 26*48. You are adding 6/2=3 and 8/4=2 to get 5 as your multiplier. The first vertical calculation is your denominator, so to speak, and launches you right into a left to right calculation. 2*4=8/5*8/48=1,248. The five is the sum of the fractions 6:2=3 and 8:4=2.

Common base or not, if you can quickly and easily add the fractions. The calculation is quite easy and is based upon the relationship of one of the vertical calculations to the proper or improper fraction it sums to. Practicing some of these ratio games will make you see the numbers and relationships as patterns that you formerly did not see at all. Don't forget to look at the ratios in both directions. The combined ratios or fractions give you a ball park feel for how big the cross product is or will be in relation to the other components, so it is a good way to check yourself as you proceed.

You should be able to look at most of these calculations and know how many digits are in your answer. Now you can also know how large your cross product is. As you go along, ask yourself how many digits in your answer? In the case of two digit multiplication, it's generally going to be four digits. How many digits in your cross product? It can go to three, but generally it will be two. It is actually easier in most cases to work with the relationships of the numbers themselves than it is to plow through the calculation, which is always an option.

As you may have guessed, you can apply aliquot parts to these calculations. If the ratio adds to a fraction that is not a whole number, it makes no difference. You can work with the fraction. In the case of a fraction made up of a whole number and a fraction, the calculation is just as simple usually as working with a whole number.

For example, look at 63*48. if you look at the fractions from right to left, you have 3:6 which is a half. The other fraction being 8:4 which is two. The first vertical calculation is 6*4=24. In this case, the cross product could be calculated as 2.5*24. It's not a difficult calculation. You could just double the 24 to get 48, halve the 24 to get 12, and put them together to get 60. Since 2 and 1/2 are reciprocals, the operation is the classic double and half you often see in regards to aliquot parts.

You can multiply 39*31 in a similar fashion. One fraction is a whole number three, and the other is its reciprocal of 1/3rd. The first vertical calculation is 3*3=9. The cross product is 3 1/3*9= 30. It looks like kind of a difficult calculation, but it really isn't. The number 27 is 3*9, and the number 3 is 1/3rd of 9. If you put 27 and 3 together, you get 30.

Use Fractions to Multiply

Notice how both of these calculations resembled the calculation of a reciprocal because the first in the last coefficients were the same. The cross product in these cases is not equal to the sum of the square of the digits, but may be equal to a multiple of the sum of square of the digits of the smallest number. In the case of 63*48, it is not apparent. In the case of 39*31 it is apparent, in that 39 is 13*3 and 13 is the reciprocal of 31. In this case, we could have computed the cross product as being three times the sum of the squares of the digits in 13 or 31. They come to 10, and 3x10=30. The point of all this is that, besides being useful to solve problems, ratios can give you some big hints about the numbers you're working with that may lead to an easy calculation when you at first thought that the problem was tough to do.

Don't forget that sometimes two fractions that don't look easy separately can sum to an easy aliquot part, or not. It really doesn't matter. Let's calculate 78*58. Here you have two fractions 7/8 and 5/8 that look messy to deal with. When you add them together, you get 12/8=1.5. Wow, that is easy to work with! The third coefficient is 8*8=64, so the cross product is 64+32=96. The components then are 35/96/64=4,524. Mixed denominators don't really affect your calculations either. If the problem had been 39*58, you would have ratios of 3/9=1/3rd and 5/8. Your last coefficient is 9*8=72. So now just add 1/3rd of 72=24 and 5/8ths of 72 which is 45. Your cross product is 69. The components and answer is 15/69/72=2,262.

John Carlin

If you think about it, any even number coupled with a five will give you a vertical calculation that is a multiple of 10. With aliquot parts, multiples of 10 are extremely easy to work with. Let's use 23x59 as an example. The first vertical product is 10. The cross product is 1.8+1.5=3.3*10=33. The math would be just as easy if the question were what's the cross product of 59*43. You would double the nine to get 18 and add 3/4 of 20 to it. So your answer is 36+15=51. Of course, this is reversible so that 95*34 would also have a cross product of 51 hooked to a trailing 20 as opposed to a leading 20. If it happens that the number associated with five is a multiple of five, and the number associated with the even number also yields a whole number, consider what happens.

Remember that you can always multiply 2 two digit numbers together by treating them as two linked single digit multiplications. For example, 54*36. You could put these together: 162/324=1,944. If the multiplier is small enough, this is not an unusual procedure. If the first digit of the multiplier was one, two, or three, I wouldn't rule this method out. 24*46 for example. With this one, you immediately have the first link. That would be 92. You know the second link is going to be twice that just by observation. So now you have to link 92/184=1,104. There was one multiplication in this process, and everything else was relational. The best way to sum this up is to say that if you look at the two numbers and you like one of the fractions and find it easy to work with, then use that one fraction relationally to help you with your two single digit multiplications. If you don't like either fraction, then separately consider adding them together, and see if that results in a fraction that is easy to use in regards to one of the vertical multiplications.

I think that we have covered ratios pretty well, and shown some of the forms that are especially useful in calculating products. We have covered situations where the numbers have a common base and end in five. We examined how this can be expanded to include numbers that have a common base and add to ten. We looked at how these forms are bidirectional. We looked at fractions or ratios and discovered how powerful they are. We saw that aliquot parts can be useful when dealing with ratios. Then we looked at a couple of special situations that involve 5 and 2, and multiplications that involve squaring numbers. Hopefully, you have some confidence now that you can square any two digit number in your head. We actually ended up with three ways to multiply two 2 digit numbers. In every case, you may be able to use fractions or ratios to help you compute the answer.

6. SYNCED RATIOS

There is one more situation I want to go over with you where fractions play a big part, and this one will astound you, and make your mental math so easy that you won't believe it! . Let's go to that problem of 39x21. In the introduction, I mentioned that the cross product of this calculation was in fact the smaller number 21. Let's look at the ratios that each of these numbers present. The number 39 in reduced fraction form is 1/3. The number 21 is the whole number two. If we add these two fractions together, we get 7/3= 21/9. The vertical multiplication we are keying on is 9x1=9. When we do the multiplication of 09* 21/9, we get 21, because the nines cancel each other out.

This calculation has a ratio such that the two fractions added together and multiplied by the product of the vertical calculation equals the smaller number. Instead of asking what causes it, let's look at the conditions that result in this kind of situation. Did you notice that the units' digits add up to 10? Did you notice that the tens' digits were one unit apart? In fact, those are the conditions that result in the ratios being synchronized as they are in the 39x21 example. Guess what? That same synchronization happens any time the units' digits add up to 10, and the tens' digits are one unit apart.

Let's look at 93*87. What would the cross product of that be? We know by calculating manually that it is (3*8)+(9*7)=24+63=87. In this kind of calculation, there really is no need to look at the fractions or to manually compute the cross product. It's already there right under your nose! All you have to do is look for the two conditions to be present. The units' digits are tens' compliments, and the tens' digits are one unit apart. Notice that I expressed the first condition as the question: "Are the units' digits tens' compliments?" I expressed it this way because I really don't want you to add the digits to see if they add up to 10. I want you to do it by observing and looking for complements as opposed to calculating. This is a subtle distinction, but it makes a difference in your perception. It's faster to observe than to calculate, so save the calculating part for when you really need it.

In the case of 29*11, we know that the vertical calculation comes to 09. The fraction that we are going to multiply by 09 is 11/9. The fraction is synchronized in that it is the same as the units' product/the multiplier. Since all these numbers are fractionally related, do you suppose that we could find other situations where some kind of fractional relationships also occur? The answer is, yes you can, but not quite like you would think.

For example, if you wanted to multiply 213*187, you would be doing a good thing to notice that 13 and 87 add up 100, and the hundreds' digits are one unit apart. In this case, you could split the number into 2/13 and 1/87. The first vertical calculation would be 02. The cross product would be 187. Now we have 02/187/ ?. The part in question would be the product of 13*87. That would be 08/31/21=1,131. So now let's put together 02/187/1131= 39,831. In the case of three digit multiplication, if the trailing two digits add up 100 and the hundreds digits are one unit apart, then your cross product is the smaller number again.

This could even work in the case of 136*64. You have to split this into 1/36 and 0/64. The vertical multiplication is zero. Your cross product is 64, and you are adding to this the product of 64/36*64=((18/48/24).=2,304)). So 64/2,304=8,704. When the units digits Into 100, and the 100s' digits are one unit apart, calculation is not much more difficult than it would be for a two digit number.

All the above is well and good, but what about when the numbers don't add up to 10? Well, that's the part where the fractional relationship starts to come in. For example let's look at 23*19. If we do a quick calculation of the cross product, we get 21. The units' digits add up to 12. So let's multiply the one in the 19 by 12. That gives us 12. How do we complete this calculation to get to 21? It looks like we have to come up with nine more units to add to the 12. The nine is right here under our nose in the last

Use Fractions to Multiply

digit of the smaller number!.

Regardless of what the units' digit sum up to, we can get the cross product by adding those unit digits multiplying by the smaller cans digit and adding the units' digit of the smaller number. Let's look at 99x89. We could do this as a fraction problem very easily. The cross product would be 17/9*81=17*9= 153. An alternative would be to add the units' digits and get 18, and multiply that by eight. In this scenario, 18*8=144. Now we add nine to that to get 153. Either method gave us the right cross product! In this case, I probably would've gone with a fractional approach. I could check my work by using the second method. The rest of the calculation in both cases is exactly the same, just a matter of linking two vertical multiplications to the cross product.

There are some other shortcuts that can be attached to the sum and difference method. If the units' digits add up to five or 15 for example, What do you think happens then? That would be an aliquot parts situation. Let's look at 83*72. Since the units add up to five, our cross product would be half of 70+2=37. That calculation would've been slowed down considerably if we tried to use fractions on it. The result is 56/37/06=5,976.

You could make the argument that though the fraction is slower to come to, it is linked to one of the vertical multiplications, and that would speed you up when you compute the answer. On the other hand, as soon as you look at that digits sum and see five, you know that the third digit of the answer is the last digit of the cross product. In effect, you know that the last two digits of the answer is 76.

If we had to multiply 89*76, we could quickly note that the units' digits add up to 15, So now our cross product is 105+6 =111. The 105 being 1.5*70. Again, this is faster than using fractions. The components then would be 56/111/54=6,764.

Now let's look at the three digit version when the last two digits of the numbers we are multiplying come to 10, 20, 30, and so on, thru 100. Let's look at 208*112. In this case, the units' digits of the multiplication add up to 20. The first vertical multiplication equals 02, the second vertical multiplication is 12*08=96. The cross product is 2/12=32. The components then would be 2/32/96=23,296. If we were going to multiply 416*314, we could do it very easily once we realized that the last two columns, or the units' digits if you will, add up to 30. Our components would be 12/(3*3/14)/(16*14)=12/104/224=13,0624.

When the last two columns of units' digits add up to a multiple of 10,   the cross product is that multiple times the smaller hundreds' digit plus the remaining two units digits of that number. Of course, when they add up to 100, then the cross product is in fact the lower number itself. When you looked at the vertical multiplication of 16 *14,   did you think to yourself that this would be 15 squared = (225 -1)=224? That calculation of 16*14 was the hardest calculation and the whole process, and it was actually quite easy.

If the units' digits don't add up to a multiple of 10, you can still do the multiplication, It is only slightly more difficult. For example,

Use Fractions to Multiply

613*529. In this case the units' digits, so to speak, add up to 42. The hundreds' digits are still one unit apart. So let's proceed and see what happens. The first vertical calculation is 6*5=30, the last vertical calculation is 13*29= 02/(12*1)+3/27=02/15/27=377. The cross product for the big calculation is (42*5)/29=21/29=239. If we put these components together we get 30/239/377=324,277. It really wasn't that difficult to calculate. When you saw that five as the lead number in the multiplier, I hope you thought that would be an easy aliquot parts 10 times the cross product divided in half. Also, did you notice how I incorporated the two digit version of this method into calculating 29*13? So this method can be used even when the units' digits don't add up to a multiple of 10. The only thing that happens when the digits don't add up to a multiple of 10 is that you have a two digit by one digit multiplication to do to get the cross product, as opposed to a single digit times a single digit.

It would appear that this is a real magical method, but it is quite limited. At first, we said it was limited because you can only use it when the units' digits add to 10 or 100. Now we have solved that problem and expanded the application to include every case, the only restriction being that the lead digits have to be one unit apart. Is there a way to expand that restriction?

The answer to that question is yes. If the difference between the lead digits were two units, then all we would have to do is multiply that difference times the units' digit, or digits of the smaller number. For example, if we wanted to multiply 39*11, what would our cross product be? We can easily calculate it to be 12. We can get that same cross product by linking 10/02=12. The 02 being two times the units' digit of the smaller number. That is all there is to it. If the question had been what is the cross product of 38*11, then we would have two adjustments to make. There would be one because the units add to nine instead of 10. Then there would be a second adjustment because the lead digits are two units apart instead of one. If we make the two adjustments, we will be linking 09/02=11. The calculation then would be 03/11/08=418.

I hope that you can see how these adjustments would progress if the difference between the lead digits were three. The adjustment would be three times the units' digit or digits if you are working with a three digit number. In the case of a difference of four or five, the progression would continue in a similar fashion. In the case of a five unit difference, aliquot parts could come into play again. For example, if you had to multiply 97*49, we could derive the cross product as 16*4/(90 divided by two)=64/45=109. The entire calculation would consist of linking 36/109/63=4,753.

I like to point out bi-directionality in every instance when I run into it. If the numbers in the calculation were turned around and the tens' digits added to 10 and the units' digits were one unit apart, would we still be able to find a cross product? In our original problem, we looked at 39*21. Let's turn those numbers around. Let's look at 93*12. We know that the cross product is still going be 21, because it does not change even when you reverse the numbers. In this

Use Fractions to Multiply

case, the 21 is still there. You just have to read it from right to left. It is the 12 read backwards!

In a sense, we just doubled the power of this method. If you look at two numbers, yet multiply together and the units' digits add to 10 and the tens digits are one unit apart, we instantly know that the cross product is the smaller number. If the situation is reversed, we can also get the cross product at a glance. In the case of a three digit number, the math is the same, but the way that you read the number backwards is slightly different.

Earlier, we computed (213*187) to be 02/187/1131=39,831 using our sum and difference method. The 13 and the 87 added to 100. If we turn that around and try to solve for 871*132, how does our cross product work out? The first vertical calculation is now 87*13=1,131. The last vertical calculation is 2*1=02. So far, all that has happened is that the first and last coefficients switched places when we reversed the numbers. The cross product if we read from right to left is 178? You have to remember though, that we're reversing the right most single digit but not the double-digit. So the cross product really is 187. The components then become 1131/187/02=114,972.

Now we have covered a super powerful additional tool to help you with the math when fractions get tough. It too is based on fractional relationships. It applies to two and three digit multiplication, and it is bidirectional too. The method is especially effective when the differences between the numbers are small, as in one, two, or three times ten, or one hundred in the case of three digit multiplication. We can see

that aliquot parts can be applied to these
multiplications. We can even have the method
looped inside the loop, so to speak, when we use
the same shortcut twice inside the same
calculation as we did when we multiplied
613*529. You formerly thought that fractions
were only for measuring bakery ingredients.
Now we have discovered that fractions are
themselves the main ingredient in multiplication!

John Carlin

7. CUBING FRACTIONS

I want to cover one more situation. It really works well for two digit numbers. Just as you can square a two digit number by using fractions, you can cube a three digit number in a similar manner. The simplest fraction to cube would probably be 11 or 12. Look at both of them. Any time you cube a two digit number, the resultant product is going to be somewhere between four and six digits. Regardless of the number of digits, you will have four coefficients or components.

In the case of 11 cubed, you could start out with the vertical calculations. One cubed is 1*1*1=1, so each vertical calculation results in one. Now we just have to fill in the middle coefficients. We know that we have two more coefficients to plug-in the middle between the vertical calculations. We also know that they have a one-to-one relation to each other. So now

Use Fractions to Multiply

we can expand what we know to be four coefficients as follows 01/01/01/01. Those middle two coefficients, though, have to be tripled since we are cubing. That gives us coefficients of 01/03/03/01=1,331, which is in fact the result when you multiply 11x11x11.

Now let's look at 12 cubed. We know that the first vertical calculation would give us 01. The last vertical calculation would be 2*2*2=8. The fraction 1/2 gives us a good description of the relationship of the two middle coefficients to either the first or the last coefficient. We can fill in the initial coefficients as 01/02/04/08. Notice how each digit is twice the adjacent digit starting from the left, or conversely half the adjacent digit starting from the right? We have used that fractional 1/2 to establish the skeleton of the problem. Now all we have to do is multiply the two inside coefficients by three. When we do that, we get 01/06/12/08=1,728 which is in fact what 12x12x12 equals. Certainly any two digit number that starts in one or ends in one can easily be cubed this way.

Let's try a few more and see what happens. If we were to cube 31, how would we do that? It's not any more difficult than cubing 13. The cube of three equals 27. The cube of one is of course one. The fractional relationship between the numbers is 1 to 3 and 3 to 1. In the case of 31 cubed, the initial skeleton would be 27/09/03/01. Now we just multiply the two middle coefficients by three to get 27/27/09/01= 29,791. If we were cubing 13. It would go like this: 01/09/27/27=2,197.

John Carlin

Let's try 24 cubed. Two cubed equals eight, and 4 cubed equals 64. The fractional relationship is 2/4=1/2. The skeleton would be 08/16/32/64. Now just multiply the two middle coefficients by three to get 08/48/96/64= 13,824.

Let's cube 93. This time, let's start on the right in writing the links, including the multiplication by three down all on one line at one time. One thing that would greatly facilitate cubing like this would be that you know all the single digit cubes one through nine. Nine cubed is quite simple. As you might guess, the digit sum is nine. 9X9x9=81*9=729. Now let's start on the left, and proceed with this together. We have 729 linked with (729x3) divided by three. The threes cancel each other out. How sweet is that? So we link 729/729=8,019. The next link is 1/9th of 729*3. 1/3*1/3* 3*729=243. again, some of the work cancels out. So this is real simple too, 81x3 equals 243. If we link 80 19/243 we get 80,433. The last step is to link this to 27. Our final answer is 804,357.

I hope this last problem shows you especially how easy it is to cube a number that involves the ratio of 1/3, or 2/3. The work is greatly facilitated by the cancellations. Somehow we have used the addition of fractions, and the multiplication of fractions to help us mentally multiply and cube numbers.

What happened to the plant in math class?

It grew cube roots! :)

Use Fractions to Multiply

8. SYNCED RATIOS FOR 3 DIGITS

If you wanted to multiply 213*187, it would be a good thing to notice that 13 and 87 add up 100, and that the hundreds' digits are one unit apart. In this case, you could split the number into 2/13 and 1/87. The first vertical calculation would be 02. The cross product would be 187. Now we have 02/187/ ?. The part in question would be the product of 13*87. That would be 08/31/21=1,131. So now let's put together 02/187/1131= 39,831. In the case of three digit multiplication, if the trailing two digits add up to 100 and the hundreds' digits are one unit apart, then your cross product is the smaller number again.

This could even work in the case of 136*64. You have to split this into 1/36 and 0/64. The vertical multiplication is zero. Your cross product is 64, and you are adding to this the product of 64/36*64=((18/48/24).=2,304)). So 64/2304=8,704. When the units' digits add to 100, and the 100s' digits are one unit apart, calculation is not much more difficult than it would be for a two digit number.

Now let's look at the three digit version when the last two digits of the numbers we are multiplying come to 10, 20, 30, and so on thru 100. Let's look at 208*112.

John Carlin

In this case, the units' digits of the multiplication add up to 20. The first vertical multiplication equals 02, and the second vertical multiplication is 12*08=96. The cross product is 2/12=32. The components then would be 2/32/96=23,296. If we were going to multiply 416*314, we could do it very easily once we realized that the last two columns, or the units' digits if you will, add up to 30. Our components would be 12/(3*3/14)/(16*14)=12/104/224=13,0624.

When the last two columns of units' digits add up to a multiple of 10, the cross product is that multiple times the smaller hundreds digit plus the remaining two units' digits of that number. Of course, when they add up to 100, then the cross product is in fact the lower number itself. When you looked at the vertical multiplication of 16 *14, did you think to yourself that this would be 15 squared = (225 -1)=224? That calculation of 16*14 was the hardest calculation and the whole process, and it was actually quite easy.

If the units' digits don't add to a multiple of 10, you can still do the multiplication, It is only slightly more difficult. For example, 613*529. In this case, the units' digits, so to speak, add up to 42. The hundreds' digits are still one unit apart. So let's proceed and see what happens. The first vertical calculation is 6*5=30, and the last vertical calculation is 13*29= 02/(12*1)+3/27=02/15/27=377. The cross product for the big calculation is (42*5)/29=21/29=239. If we put these components together, we get 30/239/377=324,277. It really wasn't that difficult to calculate. When you saw that five as the lead number in the multiplier, I hope you thought that would be an easy aliquot parts problem of 10 times the cross product divided in half. Also, did you notice how I incorporated the two digit version of this method into calculating 29*13? So this method can be used even when the units' digits don't add up to a multiple of 10. The only thing that happens when the digits don't add up to a multiple of 10 is that you have a two digit by one digit multiplication to do to get the cross product, as opposed to a single digit times a single

Use Fractions to Multiply

digit.

Earlier, we computed (213*187) to be 02/187/1131=39,831 using our sum and difference method. The 13 and the 87 added to 100. If we turn that around and try to solve for 871*132, how does our cross product work out? The first vertical calculation is now 87*13=1,131. The last vertical calculation is 2*1=02. So far, all that has happened is that the first and last coefficients switched places when we reversed the numbers. The cross product, if we read from right to left, is 178? You have to remember, though, that we're reversing the right most single digit but not the double-digit. So the cross product really is 187. The components then become 1131/187/02=114,972.

We can even have the method looped inside the loop, (nested loops so to speak), when we use the same shortcut twice inside the same calculation as we did when we multiplied 613*529. Clearly, this method can be useful not just for 3 by 3 digit multiplication but also for 3 by 2, and 4 by 2 multiplication, as we shall see.

Use Fractions to Multiply

9. 2X3 DIGITS

I own over 100 Math books. Mostly, they are about mental math and algebra. In those 100 books, I don't think I've ever seen anyone write about mixed multiplication for 2 x 3 and 4 x 2 multiplications. Everyone writes about 2 x 2 and 3 x 3 and glosses over the rest. I've certainly never seen it suggested that proportion or fractions can be used in the case of a numbers matrix that is not square. I hope you like this original material and find it useful in your multiplications.

Some of the same methods apply to 2x3 multiplication as apply to 2x2 multiplication. Now you have some options concerning how to split up a three digit number into two parts. For example, in 121*31 you could split the 121 into 12/1 and have the 31 as 3/1 to get 15/1 as your fraction. The result is 36/15/01=3,751. You could split it as 1/21 and 3/1= 3/73/21=3,751. Look carefully at that cross product. It was made up of the fraction 3/63 and 3/10. When you split the number this way, the 3/1 is actually (3/10). This can be a little tricky if you don't recognize that there is a zero to be added. Given the two ratios, I don't know why you would split these numbers in any way other than 12/1 & 3/1.

John Carlin

You could always do it as two linked single digit multiplications too. In that case, you would have 363/121=3,751. Always start out by thinking about how many digits you can expect the answer to be. It's always going to be four digits, or in most cases five digits. Then look at the relationship of the numbers to each other. Consider multiplying 842*21. If you were to do this as two single digit multiplications linked together, you would instantly know that the first link is 1,684. The second link would be 842. 1,684/842= 17,682. Using a linked single digit approach is not at all out of the question for multiplications where most of the digits involved are small. The carrying involved is therefore minimal, and you can practically do the whole procedure by observation.

Let's do several multiplications where the denominators are the same. For example, let's multiply 123*33. We can split this into 12/3 and 3/3. Now we will do the first vertical multiplication, which is 12*3 equals 36. The two fractions add up to be the whole number five. The last vertical multiplication is 3*3=9, and the cross product is 5*9=45. Our components and solution then are 36/45/09=4,059. This is as easy as 2*2 multiplication. Let's try a couple more just to make sure that we have a handle on this. When you have a common leading or trailing base, the fractional approach is the preferred method.

Let's look at 199*89. From a fractional perspective, we should split this into 19/9 and 8/9. Added together, we would then have 27/9=3 as our fraction to work with. The first component

Use Fractions to Multiply

is (19*8)=152. The cross product would be
81*3=243. The last component would (9*9)=81.
Now we just have to assemble the components
of 152/243/81. That would come to 17,711. This
is a very straightforward problem.

Let's look at (675*75). Split this into 67/5
and 7/5. The two fractions added together come
to 72/5=14.4. Our last component is 25.
14.4*25=360. Our first component is 67*7=469.
Now we have to put together 469/360/25. That
would give us 50,625.

The problem gets more difficult when we try
to split the fraction so that the last two digits
are the trailing figures. For example, let's look
at 618*23. If we split it into 6/18 and 2/3, we are
really adding when 1/3+2/3=1. Therefore, we
should have components of 12/54/54. That cross
product of 54 is actually computed as
180+36=216, instead of 36+18=54. You can get
the 180 by multiplying 3*60=180, and just add,
2*18=36 to it. As a practical problem, you would
split the three digits into a one digit and two
digit approach when the numbers fall into your
lap so to speak. You don't really want to force
the method because of the other necessary
corrections to be made.

Consider the problem of (612*93). We can
split this into 6/12 and 9/3. From a fractional
perspective, we are adding 1/2+3=3.5. Our last
component is 36. Our first component is 54 .
The cross product before adjustment would be
108+18 =126, or 3.5*36=126. The 18 is really
180, so we add 180+108=288, and we get the
correct cross product. Let's do another one.

John Carlin

This time, let's multiply (612*13). Let's split 612 into 6/12 and 13 into 1/3. The fractional result is 5/6. Therefore, the cross product is 30, and our final component is 36. The first component is 06. We need to adjust the cross product so that it is not 18+12, but rather 180+12=192.

It's not uncommon for halves and thirds to be mixed together. Hopefully, you can look at a half, and a third, and think 5/6ths right off the bat. By the same token, when you see a half and two thirds you should think 7/6ths without actually calculating it. The same thing should happen when you see one half matched with 3/8, 5/8, or 6/8 (which is three quarters). Immediately, you should think 7/8ths, 9/8ths, and 10/8ths almost reflexively. The last thing you want to do is to reinvent the wheel every time you see these fractions.

When you're doing 3 x 2 multiplication, and you keep the double-digit on the left side of your fraction, the multiplication is exactly the same as it is for a 2 x 2 situation. If you're doing a 3 x 2 multiplication, and the double-digit is on the right side of the fraction, then you have to do an additional step and adjust your proportion.

Everything that you can do with the 2 x 2 multiplication, you can do with the 3 x 2 multiplication. The same numbers that are easy to do with 2 x 2 multiplication are easy in 3 x 2 multiplication. Ones, nines, fives, thirds, and halves are all easy to work with. Numbers that are repeated are easy to work with, as they lead to common denominators and easy calculations. As you can read your fraction from left to right, or right to left, the addition of an extra digit just

Use Fractions to Multiply

affords you more opportunity to use fractions to help you multiply.

Because of the identity property, and the fact that one as a denominator means a whole number Instead of a fraction, the number one is especially easy to work with. Just as you could write a whole chapter on the number nine, you could write a whole book based on the number one and its utility in multiplying. My next book is going to be about using the identity property and the commutative property as an aid in multiplication.

Let's do some exercises so that you can develop some level of confidence, and you can see how the addition of that one more digit gives you more choices to solve the problem using fractions.

Exercises

1. 123*13

2. 159*15

3. 321*15

4. 895*23

5. 764*64

6. 832*73

7. **246*71**

8. **987*29**

9. **651*46**

10. **763*84**

11. **952*63**

12. **821*72**

13. **771*91**

14. **229*94**

15. **753*78**

16. **916*56**

17. **484*23**

18. **759*12**

19. **842*45**

20. **963*62**

21. **212*56**

22. **312*68**

23. **515*46**

24. **912*54**

Use Fractions to Multiply

25. 279*62

    Go To the next page to see how I divided these up. Use a calculator to check your work.

Answers

1. 12/3 and 1/3 = 12/39/09

2. 159 and 1/5 = 159/795   (as two linked 3 digit multiplications)

3. 321 and 1/5 = 321/165   (as two linked 3 digit multiplications)

4. 895 and 2/3 = 1790/2685 (as two linked 3 digit multiplications)

5. 76/4 and 6/4 = 456/328/16

6. 8/32 and /32 = 256/1024   (as a single and a square)

7. 24/6 and 7/1 = 168/66/06

8. 98/7 and 2/9 = 196/896/63

9. 65/1 and 4/6 = 260/394/06

10. 76/3 and 8/4 = 608/328/12

11. 95/2 and 6/3 = 570/297/06

John Carlin

12. 82/1 and 7/2 = 574/171/02

13. 77/1 and 9/1 = 693/86/01

14. 22/9 and 9/4 = 198/169/36

15. 75/3 and 7/8 = 525/621/24

16. 91/6 and 5/6 = 455/84/36

17. 48/4 and 2/3 = 96/152/12

18. 75/9 and 1/2 = 75/159/18

19. 84/2 and 4/5 = 336/428/10

20. 96/3 and 6/2 = 288/594/06

21. 2/12 and 5/6 = 10/180/72

22. 3/12 and 6/8 = 18/312/96

23. 5/15 and 4/6 = 20/360/90

24. 9/12 and 5/4 = 45/420/48

25. 27/9 and 6/2 = 162/108/18

Five out of four people don't understand jokes about fractions. :(

Use Fractions to Multiply

10. 2X4 DIGITS

We have earlier indicated that anything in the up to six digits range is fair game from a mental math and fractions approach. Two digit by four digit multiplication falls in that category. In some ways, 2 x 4 digit multiplication is easier than 2 x 3 digit multiplication. There are really only two ways to accomplish it. With smaller numbers, you can treat the multiplication as if it were two single digit multiplications of a four digit number. The only other viable option is to split the two digit number in half, and the four digit number in half.

For example, if you wanted to multiply 1234*12. This would be a fairly easy multiplication. Simply write down 1234/2468. Now link these to get 14,808. It's easy to see how well this works for small numbers. The big drawback to doing it this way is that your link is actually three digits long. That is a bit much.

John Carlin

If you look at the first half of the four digit numbers and know that you can multiply that by the two digit multiplier, you could treat the multiplication as two digit multiplications linked together. For example, if you split 1234 into the fractions (12/34)*12 you immediately could write down or mentally come up with 144/ as the first part of your multiplication. Then all that would be left to do would be multiply 34x12 and link that to the 144. 3/4+1/2= 1 &1/4. So that multiplication would result in 3/10/8= 408. 144/408 would link to be 14,808.

If you are doing this same multiplication by splitting the numbers in half, your vertical multiplications would be 12x1 and 34*2. There really isn't a ratio or fraction that you can work with in this case. Your cross product would be 120*2, plus 34x1. The components would be 12/274/68=14,808. Notice how we had to multiply the 12 by 10 to get 120. The cross product was not 24+34=58, it was 24/34 = 274.

Let's do one where there is a fractional relationship between the numbers. Try out 1224*12. We can split this into 12/24 and 1/2 as our fractions. Together, they add up to one that means that our components are 12/48/48. But the cross product 48 has to be split into 24/24=264. Now let's put those components together as 12/264/48= 14,688. This is a very easy calculation because the cross product is the same as third coefficient, and in that each part is one half, the cross product itself is split in half and rejoined. Thus 48 became 24/24. Let's try one that's a little more difficult.

Use Fractions to Multiply

Let's look at 1296*78. I would split these numbers into 12/96 and 7/8. In this example, the fractions add up to one, 12/96=1/8 and combined with the 7/8ths, you have a whole number of one as the result. That means that the components are 84/768/768. The cross product of 768 is really 96/672=10,272. So now, we are combining 84/96/672/768=84/10,272/768= 943,488. There really are four components to the multiplication, but with the use of fractions, you can start it out as three components, and then adjust or split the cross product into two parts. In some cases, it is really easy to do, and in others, it is a little more difficult. For sure, you can now appreciate all the emphasis that we put on multiplying two digit by one digit multiplication. Without a good foundation there, the whole multiplication quickly degenerates into a difficult problem. Some of the problems you just have to power through using four consecutive 2x1 multiplications, and some will have a fraction that will be easier to solve. Try the following exercises.

1. 9613*23

2. 1456*41

3. 1729*34

4. 3648*24

5. 1767*59

John Carlin

6. **8119*38**

7. **8642*82**

8. **6137*23**

9. **6874*57**

10. **8891*39**

Use Fractions to Multiply

Answers

1. 2208/299=221,099

2. 574/2296=59,969

3. 578/06/35/36=58,786

4. 06/24/24/08/32/32=864/1152=87,552

5. 85/153/335/603=104,253

6. 243/648/57/152=308,522

7. 64/64/12/32/24/04=7052/3444= 708,644

8. 1403/06/23/21=1403/851=141,151

9. 340/476/370/469=391,818

10. 24/96/72/27/84/09=346,749

11. 3X3 DIGITS

With 3 x 3 multiplication, there are four ways that you can split the matrix. For example, if you wanted to multiply 123*456, you could split your numbers into 1/23, and 4/56, or 12/3 and 45/6, or 12/3 and 4/56, or 1/23 and 45/6. Additionally, you could do your multiplicand times each digit. In other words, 123*4, linked with 123*5, and 123*6. None of these would be really tough multiplications. In this case, you would be linking 492/615/738= 56,088.

Your choice of how to split the multiplier multiplicand into fractions would depend upon what made the most sense to you. Clearly, you would choose a split that would make the most number of fractions resulting in easy calculations. How would you split up 123*456? Personally, I probably would split this one into 4/56 and 1/23. Let's see how that works out. The first vertical multiplication would be 04. Then I would have cross products that would be the sum of 23 *4 and 1*56. So now we would have to link 04/148. The second part of the calculation would be the addition of the vertical 56*23. The fractions add to 1 and 1/2. The complete multiplication results in 10/27/18=1,288. So now

Use Fractions to Multiply

we have to link up 04/148/1,288=56,088.

There are no adjustments to be made when the fractions are split, so there is a direct correlation between the number of digits in the multiplier and in the multiplicand. If you have the same number of digits digits on the left or the right side in both the multiplier and the multiplicand, no adjustment is necessary.

As you might guess, you do have to make an adjustment when you split the fractions so that one has a two digit numerator and the other a two digit denominator. Let's take a look at an example involving that. If the earlier problem had been to find the product of 123*324, I'm a bit inclined to split it into 12/3 and 3/24. My first vertical calculation would be 12*3=36. My fraction adds up to 4 and 1/8. My last vertical calculation would be 3*24 = 72. Now my cross product should be equal to 72 times 4 and 1/8th. The cross product is equal to 288+9=297. That cross product has to be adjusted though. It is really 288+90= 378. At this point, this is not an unfamiliar adjustment. We still had three components that were fairly easy to calculate an adjustment. It was fairly simple, and we had our answer.

The key to all these multiplications is to look first, and split the fractions in the way that make for the easiest calculation. There are always numbers that you can build upon. All the digits one through five are pretty easy to work with. They are the easy fractions halves, thirds, fourths, and fifths. Anything that adds up to a whole number is very simple to work with.

ABOUT THE AUTHOR

John Carlin is a long time amateur mathematician who lives in Apple Valley, Minnesota. He enjoys the beauty, symmetry, and immutability of math, and hopes you will come to appreciate it as well. He is a graduate of the U.S. Merchant Marine Academy in Kings Point, N.Y., and has an MBA from the University of Minnesota.

www.ingramcontent.com/pod-product-compliance
Lightning Source LLC
Chambersburg PA
CBHW071302170526
45165CB00003B/1385